Developing Skills with

# TABLES
# AND GRAPHS

Book A
Grades 3-5

Elaine C. Murphy

DALE
SEYMOUR
PUBLICATIONS
P.O. BOX 10888
PALO ALTO, CA 94303

Editor: E.M. de Silva
Production Coordinator: Ruth Cottrell
Illustrator: Pat Rogondino
Cover Designer: John Edeen
Compositor: WB Associates

ISBN 0-86651-014-1

Order Number DS01171

cdefghi-MA-876543

DALE
SEYMOUR
PUBLICATIONS
P.O. BOX 10888
PALO ALTO, CA 94303

# CONTENTS

## TALLIES

## TABLES

## PICTURE GRAPHS

## BAR GRAPHS

## INTRODUCING CIRCLE GRAPHS

## INTRODUCING LINE GRAPHS

# INTRODUCTION

This is one of two books on Developing Skills with Tables and Graphs. It contains 48 worksheets that you can duplicate and use with your students for individual work or class projects. The activities in these books were designed to help students at grade levels 3-5 (Book A) and 6-8 (Book B) develop their skills in reading, using, and making tables and graphs.

## ABOUT THIS BOOK

**Four Major Objectives**

The worksheets in this book (Book A) concentrate on tallies, tables, bar graphs, picture graphs, and include some work on circle graphs and line graphs. The exercises are designed to help you teach students

- to recognize different parts of tables and graphs such as titles, headings, and keys.
- to find specific facts in various tables and graphs.
- to make their own tables and graphs.
- to interpret and use information presented in tables and graphs.

**Non-threatening Math**

The exercises have been written specifically to help you focus on developing skills with tables and graphs. The size and type of numbers students encounter sometimes seems to overwhelm their thinking. In this book, students are required to use only basic arithmetic operations with which they are familiar and numbers with which they are comfortable. As a result, students should not find the computation work threatening and will be able to concentrate their efforts on learning the special skills needed to use tables and graphs effectively.

**Focus on Organizational Skills**

Responses to test items administered by the most recent National Assessment of Educational Progress indicate that, in many cases, students' ability to read tables and graphs is superficial. They can find facts, but have difficulty responding correctly to questions that require them to go beyond a direct reading of the information. In order to alleviate these problems, the activities in this book pose questions which require students to reconstruct how data in a graph has been organized, interpret the meaning of keys, and focus their attention on titles, headings, and labels. In addition, students practice problem-solving tasks, learning to process more than one piece of information by learning to recognize whether there is enough information to answer a question, making comparisons and predictions, and drawing conclusions from facts.

# USING THIS BOOK

**Pick-and-Choose Lessons**

Most pages are written so that individual students can work independently, giving you maximum flexibility in organizing your class. You can assign activities to individuals or small groups, or you can use them for a class lesson. You can select single exercises as needed or present a combination of worksheets as a single unit of study.

**Different Kinds of Activities**

There is no required order for presenting the exercises in this book. To make it easier to locate worksheets you need, the pages are organized into sections by type of graph. The first worksheet of each section presents the main features of the table or graph in that section. The other activities within a section are loosely sequenced by skills such as finding facts, using them, completing a graph or making one. For example, one page requires students to take a poll and record the results. In another activity, students must determine from a bus schedule whether or not a person will be late to school. In a few cases, extra materials such as dice are needed. (The materials required are indicated at the top of the page.) The last activity is a review page, giving you a chance to check up on students' basic knowledge of tables and graphs.

**Three Levels of Difficulty**

To give you a better idea of what you can expect, the contents shows one, two, or three dots to indicate level of difficulty, from easiest to most difficult. These indications should be considered only as a guide because the difficulty of the exercises depends largely on the backgrounds of your students and the skills they already have acquired.

**Recording Forms**

Some pages suggest follow-up activities for which students must create their own graphs. Other pages may act as a natural catalyst for new ideas and graphing activities. For your convenience, two different blank forms and a page of cut-out coins which can be pasted on the forms are provided.

**Beyond These Pages**

Being able to read and interpret graphs, charts, and tables is an essential skill for making many everyday decisions. Draw upon different situations both in and out of math class to sharpen students' graphing skills and to point out the usefulness of presenting facts in an organized way. Expose students to a variety of graphs and tables used in real life situations — menus, TV schedules, sports statistics, and so on. Encourage students to collect their own examples and discuss the many different ways information can be organized, considering the advantages and disadvantages of each. Teach students to make and use tables and graphs to help solve problems. With time and practice, your students' skills with tables and graphs will be useful tools for making information more meaningful and manageable.

Developing Skills with

# TABLES
# AND GRAPHS

# TALLIES

You make a tally to help keep count.

There are seventeen words in this sentence and the next. The tally at the right shows seventeen.

You make one tally mark for each thing you count.

/

The fifth tally mark goes across.

卌

Complete each of the following.
Write your answers on the blanks.

1. How many tally marks? Count by fives.

卌 _____          卌 ||| _____

卌 卌 _____          卌 卌 ||| _____

卌 卌 卌 _____          卌 卌 卌 ||| _____

卌 卌 卌 卌 _____          卌 卌 卌 卌 ||| _____

2. Make tally marks for each number.

3 _____          18 _____

5 _____          25 _____

12 _____          26 _____

7 _____          21 _____

1

# 👁 COLOR

Lori counted different eye colors in her class.
She made a tally to show the results.

EYE COLOR

| color | tally |
|-------|-------|
| blue | ||||| || |
| brown | ||||| ||||| |
| brown | ||||| |||| |
| green | ||| |

**Answer each question.**

1. What colors did Lori count? Name them.

   _____    _____    _____

2. Copy the tally marks next to blue. _____
   How many marks did you copy? _____

3. How many students have blue eyes? _____

4. How many marks are next to brown? _____

5. How many students have brown eyes? _____

6. How many marks are next to green? _____

7. How many students have green eyes? _____

8. How many students are in Lori's class? _____

2

# DIAL-A-DIGIT

Teya's phone number is 664-2162.
She made a tally chart about
her phone number.

| digit | number of times |
|-------|-----------------|
| 1 | I |
| 2 | II |
| 4 | I |
| 6 | III |

1. Yukio's phone number is 615-5513.
   Complete the tally chart about Yukio's
   phone number. Use tally marks.

| digit | number of times |
|-------|-----------------|
| 1 | II |
| 3 | |
| 5 | |
| 6 | |

2. Angel's phone number is 926-7557.
   Make a tally chart about Angel's phone number.

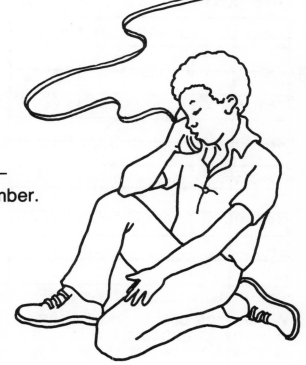

3. Write a phone number on the blanks.

   ___ ___ ___ - ___ ___ ___ ___

   Make a tally chart about that phone number.

Name _____

# CALENDAR CHARTS

You can make tally charts about the numbers in a calendar.

**1.** Complete the chart about the calendar.

| digit | number of times |
|-------|-----------------|
| 0 | /// |
| 1 | ||||  ||||  //// |
| 2 | |
| 3 | |
| 4 | |
| 5 | |
| 6 | |
| 7 | |
| 8 | |
| 9 | |

**2.** Which digit occurs most often in May? _____

**3.** Make a chart about July.

| digit | number of times |
|-------|-----------------|
|  |  |
|  |  |
|  |  |
|  |  |
|  |  |
|  |  |
|  |  |
|  |  |
|  |  |
|  |  |

**4.** Which digit occurs most often in July? _____

**5.** Is your answer the same as for May? _____

# SOCK IT TO ME

SOCKS STUDENTS WORE IN
MR. QUESADA'S CLASS YESTERDAY

| sock color | number of students |
|------------|--------------------|
| white | ⅢⅠ ⅢⅠ Ⅱ |
| blue | ⅢⅠ |
| brown | ⅢⅠ Ⅱ |
| black | Ⅱ |
| green | Ⅲ |
| other | ⅢⅠ |

**Answer each question about socks in Mr. Quesada's class.**

1. How many students wore white socks? _____

2. How many students wore black socks? _____

3. How many students wore green socks? _____

4. Which color did more students wear, blue or brown? _____

5. Which color did more students wear, blue or green? _____

6. Which color did most students wear? _____

7. How many students were counted? _____

8. How many students did *not* wear white socks? _____

9. How many students did *not* wear brown socks? _____

5

# HAIRY RESULTS

**Make your tally below.**

HAIR   COLOR

| color | tally |
|---|---|
| black | |
| brown | |
| blonde | |
| red | |

**Answer each question about your tally.**

1. How many have black hair? _____

2. How many have brown hair? _____

3. Are there more kids with blonde hair or with red hair? _____

4. Which hair color did you count most often? _____

5. Which hair color did you count least often? _____

# TABLES

**Tables** are used to show facts.  Each part of a table has a meaning.

Most tables have a **title**.

1. The title of this table is

   _____

   _____ .

| NUMBER OF PLAYERS ON TEAM ||
| game | players |
| --- | --- |
| baseball | 9 |
| soccer | 11 |
| basketball | 5 |
| football | 11 |
| ice hockey | 6 |

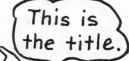
This is the title.

Tables have **headings**.

2. This table has two

   headings across the top.

   They are _____

   and _____ .

| FREE THROW CONTEST ||
| name | shots made |
| --- | --- |
| Inez | 8 |
| Hori | 3 |
| Albert | 5 |
| Elaine | 7 |

These are the headings.

Tables have **facts**.  The facts in this table are the names of the teams and the number of points they scored.

3. The 4 team names are

   _____, _____,

   _____, and

   _____ .

| SCORES IN RALLY ||
| team | points |
| --- | --- |
| Lions | 7 |
| Eagles | 8 |
| Bears | 10 |
| Roosters | 8 |

4. The 4 scores are _____,

   _____, _____, and _____ .

# READING A CALENDAR

Irene writes special days on her calendar.

| FEBRUARY | | | | | | |
|---|---|---|---|---|---|---|
| Sun. | Mon. | Tues. | Wed. | Thurs. | Fri. | Sat. |
| | | | | | 1 | 2 |
| 3 | 4 Dentist | 5 | 6 | 7 | 8 | Mom's 9 birthday |
| 10 | 11 | Lincoln's 12 birthday | 13 | Valentine's Day 14 | 15 | 16 |
| 17 | 18 | 19 | 20 | 21 | Washington's 22 birthday | 23 |
| 24 | 25 | Trip to 26 museum | 27 | 28 | 29 | |

**Answer each question about Irene's calendar.**

1. On what day of the week did Irene visit the dentist? _____

2. On what day of the week was Washington's birthday? _____

3. What happened on February 9? _____

4. What happened on February 26? _____

5. Irene was sick 2 days after Valentine's day.

   What was the date? _____

6. Irene took a permission slip to school a week before she

   went to the museum. What was the date? _____

7. The last day of January was on what day of the week? _____

8. Irene has a music lesson every Friday.

   How many music lessons did she have in February? _____

8

# GETTING BETTER

This table shows facts about sit-ups some kids did for two weeks.

### SIT-UP RECORD

| Name | First Week | Second Week |
|------|-----------|-------------|
| Ray | 25 | 40 |
| Lee | 20 | 20 |
| Gus | 15 | 30 |
| Anna | 15 | 40 |
| Mori | 20 | 30 |

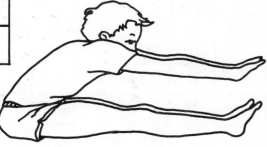

**Answer each question about the table.**

1. How many sit-ups did Ray do the first week? _____

2. How many sit-ups did Ray do the second week? _____

3. How many sit-ups did Lee do the second week? _____

4. How many sit-ups did Mori do the first week? _____

5. In the first week who did more sit-ups, Lee or Anna? _____

6. In the second week who did more sit-ups, Lee or Anna? _____

7. Who did the most sit-ups the first week? _____

8. Who did the most sit-ups the second week? _____

Find out how many more sit-ups each person did the second week than the first. Then decide who improved the most.

Write your answer here. _____

Name _____

# EXERCISES

Jackie did 10 jumping jacks, 15 sit-ups, and 12 touch-toes.
Otis did 20 jumping jacks, 20 sit-ups, and 20 touch-toes.
Gretchen did 25 jumping jacks, 10 sit-ups, and 15 touch-toes.
Ben did 15 jumping jacks, 25 sit-ups, and 10 touch-toes.

**Fill in the table to show the exercises they did.**

- Put the kids' names along the side.
- Put the exercises across the top.
- Fill in the numbers.

TITLE: _____

| Kid's Names | | | |
|---|---|---|---|
| | | | |
| | | | |
| | | | |
| | | | |

**Answer each question. Use your table.**

1. How many jumping jacks did they do in all? _____
2. How many sit-ups did they do in all? _____
3. How many touch-toes did they do in all? _____
4. How many sit-ups did Jackie and Otis do in all? _____
5. How many touch-toes did Gretchen and Ben do in all? _____
6. How many jumping jacks did Otis and Ben do in all? _____
7. Who did the most jumping jacks? _____
8. Who did the most sit-ups? _____
9. Who did the most touch-toes? _____

© 1981 by Dale Seymour Publications          10

Name _____

# AM DAILY * Monday, May 5

Angel is making a table to show temperatures.
He started to find the change from high to low.

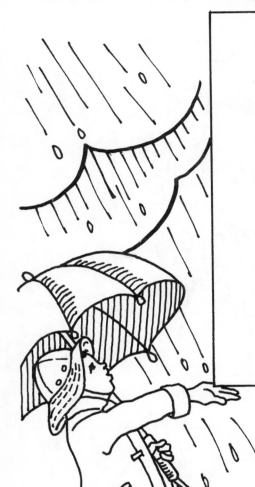

| TEMPERATURES ACROSS THE NATION | | | |
|---|---|---|---|
| City | High | Low | Change |
| Anchorage | 54° | 30° | _24°_ |
| Buffalo | 66° | 44° | _22°_ |
| Casper | 65° | 35° | _____ |
| Detroit | 82° | 45° | _____ |
| Houston | 85° | 70° | _____ |
| Las Vegas | 90° | 60° | _____ |
| Memphis | 81° | 53° | _____ |
| Philadelphia | 79° | 58° | _____ |
| San Juan | 95° | 79° | _____ |

**Complete the table for Angel.**
**Then answer these questions about the table.**

1. Which city had the highest high? _____

2. Which city had the lowest low? _____

3. Which city had the greatest change in temperature? _____

4. Which city had the least change in temperature? _____

5. Which city was warmer, Buffalo or San Juan? _____

6. Which city was colder, Houston or Anchorage? _____

Name _____

# HORROR STORIES

Hans bought a book of horror stories.
Here is the table of contents from his book.

### CONTENTS

**Answer each question about Hans' book.**

1. How many stories are in the book? _____

2. What is the title of the first story? _____
   _____

3. What is the title of the last story? _____
   _____

4. *Dr. Zooey* starts on what page? _____

5. The third story starts on what page? _____

6. The fourth story starts on what page? _____

7. *The Tick-Tock Man* ends on what page? _____

8. *The Tick-Tock Man* is how many pages long? _____

9. *From Roses to Onions* ends on what page? _____

10. *From Roses to Onions* is how many pages long? _____

# RIDING THE BUS

## BUS SCHEDULE

| Bus | Arrives at Willow St. | Arrives at Bird Ave. | Arrives at Wolfe Rd. | Arrives at High St. |
|---|---|---|---|---|
| First Bus | 7:36 AM | 7:45 AM | 7:54 AM | 7:59 AM |
| Second Bus | 8:09 AM | 8:18 AM | 8:27 AM | 8:32 AM |
| Third Bus | 8:38 AM | 8:47 AM | 8:56 AM | 9:01 AM |

Katie rides the bus to school. She gets on at Willow Street. Her school is at High Street.

**Use the bus schedule to answer each question.**

1. Suppose Katie takes the first bus.

   When will she get to school? _____

2. Suppose Katie takes the second bus.

   When will she get to school? _____

3. Suppose Katie takes the third bus.

   When will she get to school? _____

4. School starts at 8:40 AM.

   If Katie takes the first bus, will she be late? _____

   If Katie takes the second bus, will she be late? _____

   If Katie takes the third bus, will she be late? _____

5. How many minutes does it take the first bus to get from Willow Street to High Street? _____

13

Name _____

# TIME TRIALS

## How Long It Takes

| activity | Rika | Drew | Yvette | Manny |
|---|---|---|---|---|
| hop 10 times | 5 sec | 13 sec | 10 sec | 6 sec |
| clap hands 20 times | 3 sec | 4 sec | 5 sec | 4 sec |
| write name 5 times | 25 sec | 19 sec | 21 sec | 20 sec |
| count by 2's to 100 | 25 sec | 20 sec | 18 sec | 28 sec |
| blink eyes 15 times | 7 sec | 8 sec | 8 sec | 9 sec |
| count backwards from 30 | 18 sec | 18 sec | 16 sec | 15 sec |
| say hello 20 times | 8 sec | 10 sec | 10 sec | 11 sec |
| smile, then frown 10 times | 14 sec | 12 sec | 15 sec | 10 sec |

**Do you believe the following? Circle *yes* or *no*.**

1. Rika can hop 10 times in 5 seconds.      yes   no

2. Drew can write his name 5 times in 5 seconds.      yes   no

3. Yvette can count backwards from 30 in 6 seconds.      yes   no

4. Manny can smile, then frown 10 times in 25 seconds.      yes   no

5. Rika can hop 20 times in 12 seconds.      yes   no

6. Manny can hop 20 times in 12 seconds.      yes   no

7. Drew can count by 2's to 50 in 10 seconds.      yes   no

8. Yvette can say hello 10 times in 5 seconds.      yes   no

9. Manny can clap his hands 20 times in 2 seconds.      yes   no

10. Rika can blink her eyes 15 times,
    then count backwards from 30 in 28 seconds.      yes   no

14

Use a tape measure.                                  Name _____

# DO YOU MEASURE UP?

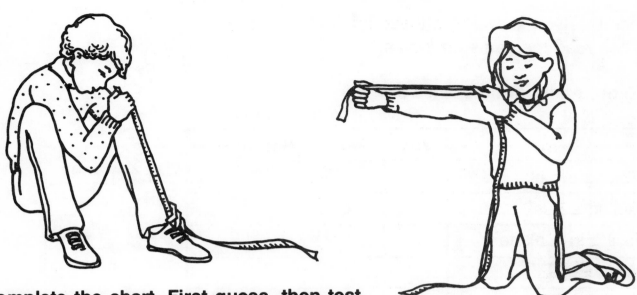

**Complete the chart. First guess, then test.**
**Use a tape measure. Make up more by yourself.**

| Parts of Me | Guess Measurement | Test Measurement |
|---|---|---|
| 1. my height | | |
| 2. my widest smile | | |
| 3. my head (top to chin) | | |
| 4. my longest arm | | |
| 5. my ankle | | |
| 6. my thumb | | |
| 7. | | |
| 8. | | |
| 9. | | |
| 10. | | |

Are you a good guesser?     yes     no

15

# HO HUM

Ask kids you know, "Are these boring?"
Then read the items on the list.

**Keep a tally of their answers.**

| Item | Yes | No | Not sure |
|---|---|---|---|
| Taking a bath | | | |
| Riding a bus | | | |
| Flying in a plane | | | |
| Watching TV | | | |
| Cleaning house | | | |
| Reading | | | |
| Rollerskating | | | |
| Visiting friends | | | |
| Going out to eat | | | |
| Being alone | | | |
| Answering questions | | | |
| Getting dressed | | | |
| Riding bikes | | | |
| Swimming | | | |
| Eating | | | |
| Shopping | | | |
| Memorizing | | | |

**Answer each question about your tally.**

1. Which item bored kids the most? _____

2. Which item bored kids the least? _____

3. How many kids answered your questions? _____

Name _____

# PICTURE GRAPHS

**Picture graphs** use pictures to compare facts.
Each part of a picture graph tells you something.

Picture graphs have **titles**.

1.  The title of this picture graph is

    _____ .

### MILK SALES

| | |
|---|---|
| Monday | 🥛 🥛 🥛 🥛 |
| Tuesday | 🥛 🥛 🥛 |
| Wednesday | 🥛 🥛 🥛 🥛 |
| Thursday | 🥛 🥛 |

🥛 means 4 cartons

Picture graphs have **headings**.

2.  The headings in this graph
    are names of people.

    The names are _____ ,

    _____ , _____ ,

    and _____ .

### TICKETS SOLD

| | |
|---|---|
| Dale | ▭ ▭ |
| J.B. | ▭ ▭ ▭ ◪ |
| Anna | ▭ |
| Gail | ▭ ▭ ▭ |

▭ means 2 tickets

Picture graphs have **pictures**.
The pictures tell you how many.

3.  The pictures in this graph

    are little _____ .

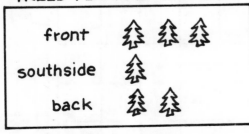

### TREES PLANTED AT SCHOOL

| | |
|---|---|
| front | 🌲 🌲 🌲 |
| southside | 🌲 |
| back | 🌲 🌲 |

🌲 means 1 tree

Picture graphs have a **key**. The key
tells how many each picture stands for.

4.  In this graph  means

    _____ .

### MITTENS IN LOST & FOUND

| | |
|---|---|
| Monday | 🧤 🧤 |
| Tuesday | 🧤 🧤 |
| Wednesday | 🧤 🧤 🧤 |
| Thursday | 🧤 🧤 |
| Friday | 🧤 |

🧤 means 2 mittens

17

# BIRTHDAY BREAK

Luis counted birthdays in his class.

He used a tally chart.

| NUMBER OF BIRTHDAYS | |
| --- | --- |
| day of week | tally |
| Sunday | IIII |
| Monday | ﭤﺖ |
| Tuesday | IIII |
| Wednesday | II |
| Thursday | ﭤﺖ II |
| Friday | ﭤﺖ |
| Saturday | III |

**Complete the problems about Luis' chart.**

**1.** Which day has the most birthdays? _____

**2.** Which day has the fewest birthdays? _____

**3.** Luis colored 4 squares to show 4 birthdays on Sunday.

Color squares to show how many birthdays on the other days. Use a different color for each day.

### NUMBER OF BIRTHDAYS

| | | | | | | | | | |
| --- | --- | --- | --- | --- | --- | --- | --- | --- | --- |
| Sunday | ■ | ■ | ■ | ■ | | | | | |
| Monday | | | | | | | | | |
| Tuesday | | | | | | | | | |
| Wednesday | | | | | | | | | |
| Thursday | | | | | | | | | |
| Friday | | | | | | | | | |
| Saturday | | | | | | | | | |

Name _____

# SCARY SCORE

David asked kids in
his class which thing
is scariest. He put
one X for each answer.

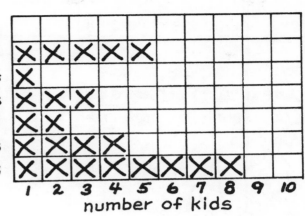

| | 1 | 2 | 3 | 4 | 5 | 6 | 7 | 8 | 9 | 10 |
|---|---|---|---|---|---|---|---|---|---|---|
| dinosaurs | | | | | | | | | | |
| Dracula | X | X | X | X | X | | | | | |
| dogs | X | | | | | | | | | |
| shadows | X | X | X | | | | | | | |
| spiders | X | X | | | | | | | | |
| snakes | X | X | X | X | | | | | | |
| monsters | X | X | X | X | X | X | X | X | | |

number of kids

## Answer the questions about David's graph.

1. How many kids said Dracula? _____

2. How many kids said dogs? _____

3. How many kids said shadows? _____

4. How many kids said monsters? _____

5. How many kids said spiders? _____

6. How many kids said snakes? _____

7. How many kids said dinosaurs? _____

8. How many more kids said shadows than said dogs? _____

9. How many more kids said snakes than said dogs? _____

10. How many fewer kids said snakes than said monsters? _____

11. How many fewer kids said dinosaurs than said Dracula? _____

12. How many answers did David get? _____

Write three more questions about David's graph.
Then answer your questions.

_____

_____

_____

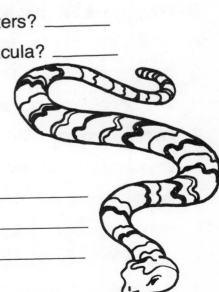

Name _____

# COLOR ME

Daniel helped the children in Mr. Greenberg's kindergarten class make a picture graph. Each child put up one square to show a favorite color.

**FAVORITE COLORS**

number of children: 8 7 6 5 4 3 2 1

red  orange  yellow  green  blue  purple  white  black  brown

color

**Answer each question about Daniel's graph.**

1. What is the title of the picture graph? _____

2. How many colors are listed in the graph? _____

3. What does ■ mean on the graph? _____

4. Which color is the most popular? _____

5. Which color is the least favorite? _____

6. Which color do more children like, yellow or green? _____

7. Which color do more children like, red or black? _____

8. Which colors do two children like? _____

9. Which colors do three children like? _____

10. Which colors do four children like? _____

11. How many children put squares on the picture graph? _____

12. How many children did *not* choose blue? _____

20

Name _____

# HOW WIDE?

Juana and Harold made a picture graph about their classroom.

WIDTH OF FURNITURE IN CLASS

| furniture | width |
|-----------|-------|
| desk | 🥾 🥾 🥾 |
| chair | 🥾 🥾 |
| chalkboard | 🥾 🥾 🥾 🥾 🥾 🥾 🥾 |
| door | 🥾 🥾 🥾 |

🥾 means 1 foot

**Complete each problem about the picture graph.**

1. How many feet wide?

    desk _3 feet____

    chair _____

    chalkboard _____

    door _____

2. Name the furniture.

    3 feet _desk , door____

    2 feet _____

    7 feet _____

3. Which is wider?

    desk or chair _____

    chair or chalkboard _____

    chalkboard or door _____

    desk or chalkboard _____

4. Which is narrower?

    desk or chair_____

    chair or chalkboard _____

    door or chalkboard _____

    chair or door _____

5. Which two pieces of furniture have the same width? _____

6. List the furniture in order from widest to narrowest.

_____     _____     _____     _____

7. A teacher's desk is 5 feet wide. Which picture shows its width?
   Circle the correct picture.

21

Name _____

# YOU GET WHAT YOU PAY FOR

This picture graph was in the school cafeteria yesterday.

**COST OF LUNCH FOOD**

| <u>Food</u> | <u>Cost</u> |
|---|---|
| milk | (10¢) (10¢) (10¢) (10¢) |
| apple | (10¢) (10¢) (10¢) (10¢) |
| sandwich | (10¢) (10¢) (10¢) (10¢) (10¢) (10¢) (10¢) |
| hamburger | (10¢) (10¢) (10¢) (10¢) (10¢) (10¢) (10¢) (10¢) |
| soup | (10¢) (10¢) (10¢) |

(10¢) means 10 cents

**Answer each question about the picture graph.**

1. How much is one apple?
   Circle the correct picture.

   (10¢) (10¢) (10¢) (10¢) (10¢)    (10¢) (10¢) (10¢) (10¢) (10¢) (10¢)    (10¢) (10¢) (10¢) (10¢)

2. How much is one milk plus one apple?
   Circle the correct picture.

   (10¢) (10¢) (10¢) (10¢) (10¢) (10¢) (10¢) (10¢) (10¢)    (10¢) (10¢) (10¢) (10¢) (10¢) (10¢) (10¢)    (10¢) (10¢) (10¢) (10¢) (10¢) (10¢) (10¢) (10¢)

3. How much are two milks?
   Circle the correct amount.

   10¢    20¢    30¢    40¢    50¢    60¢    70¢    80¢    90¢

4. Jackie bought two sandwiches and one milk.

   How much did she pay? _____

5. Joey bought two milks, one hamburger, and one soup.

   He gave the cashier two dollars.

   What was his change? _____

22

Name _____

# AT THE TOTAL TOY STORE

Jason and Becka saw this sign on a store window.

| YO-YO | PADDLEBALL | JOKE BOOK | JACKS and BALL |
|---|---|---|---|
| 70 cents | 60 cents | 30 cents | 30 cents |

**Complete the picture graph below. Then answer the questions.**

TITLE: _____

|  |  |  |  |  |  |  |  |  |
|---|---|---|---|---|---|---|---|---|
| yo-yo |  |  |  |  |  |  |  |  |
| paddleball |  |  |  |  |  |  |  |  |
| joke book |  |  |  |  |  |  |  |  |
| jacks and ball | 10¢ | 10¢ | 10¢ |  |  |  |  |  |

 means 10 cents

1. What is the title of your graph? _____

2. What does (10¢) mean? _____

3. Silly Putty costs 80 cents. How many (10¢) show the cost? _____

4. Suppose a jawbreaker costs 5 cents.

   Draw a picture using (10¢) to show the cost. _____

5. Suppose a little Frisbee costs 35 cents.

   Draw a picture using (10¢) to show the cost. _____

# JOLLY'S JUNGLE OF JUNK

Judy made a picture graph about Jolly's Store.

Prices in Jolly's Store

⑤⓪ means 50 cents

**Circle the correct answer.**

1. Which one of the following is *not* shown on the graph?

   **a.** the cost of a poster

   **b.** the cost of marbles

   **c.** the cost of five items

   **d.** the cost of balloons

2. Which one of the following costs less than one dollar?

   **a.** poster

   **b.** plastic bugs

   **c.** dice

   **d.** comic book

3. What is the cost of three comic books?

   **a.** $0.50

   **b.** $1.00

   **c.** $1.50

   **d.** $3.00

4. What is the cost of two posters and one set of balloons?

   **a.** $2.00

   **b.** $3.50

   **c.** $6.50

   **d.** $7.00

5. Funny teeth cost $1.50. Which picture shows the cost of funny teeth?

   **a.** ⑤⓪ ⑤⓪ ⑤⓪

   **b.** ⑤⓪ ⑤⓪ ⑤⓪ ⑤⓪

   **c.** ⑤⓪ ⑤⓪ ⑤⓪ ⑤⓪ ⑤⓪

   **d.** ⑤⓪ ⑤⓪ ⑤⓪ ⑤⓪ ⑤⓪ ⑤⓪

6. Which two things cost the same?

   **a.** posters and comic books

   **b.** posters and dice

   **c.** plastic bugs and balloons

   **d.** comic books and balloons

# TOWN TOTALS

Imogene saw this picture graph in her new book.

POPULATION IN CITIES

Uptown

Downtown

Intown

Lowtown

Hitown

Overtown

means one thousand people

## Answer each question about the graph.

1. Which city has more people, Uptown or Hitown? _____

2. How many more people are in Downtown than in Overtown? _____

3. Which two cities have the same population? _____

4. Which city has the fewest people? _____

5. What is the population of Overtown? _____

6. What is the difference in population between Intown and Downtown? _____

25

# FAIR WEATHER FUN

AMERICANS' OUTDOOR FUN LAST YEAR

Picnicing    ☺ ☺ ☺ ☺ ☺ ☺ ☺ ☺

Driving      ☺ ☺ ☺ ☺ ☺ ☺ ☺ ☺

Swimming     ☺ ☺ ☺ ☺ ☺

Playing games ☺ ☺ ☺ ◖

Boating      ☺ ☺ ◖

☺ means 20 million people

**Answer each question about the graph.**

**1.** The graph tells about what kinds of outdoor fun?

_____

**2.** How many people do the pictures stand for?

☺          20,000,000

☺ ☺        _____

☺ ☺ ☺      _____

☺ ☺ ☺ ☺ ☺ ☺   _____

◖          _____

☺ ☺ ◖      _____

**3.** How many people went on picnics last year? _____

**4.** How many people went swimming last year? _____

**5.** How many people went on boats last year? _____

Ask your friends about their outdoor fun.
Make a graph to tell about it.

26

Name _____

# BAR GRAPHS

**Bar graphs** use bars to show how much or how many.

Bar graphs have **titles** and **labels**.

1.  The title of this bar graph is

    _____.

2.  The labels along the side of this
    graph are names of fish. They are

    _____, _____,

    _____, and _____.

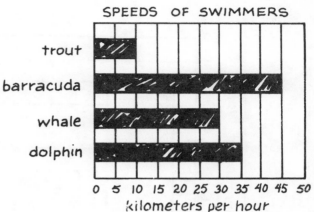

SPEEDS OF SWIMMERS

trout / barracuda / whale / dolphin

0  5  10  15  20  25  30  35  40  45  50
kilometers per hour

3.  The labels across the bottom
    of this graph are kilometers per
    hour and numbers. The numbers
    show how many kilometers per hour.

    They start at _____ and end at _____.

Bar graphs use **bars** to show the facts.

4.  Sometimes the bars go
    up-and-down. How many
    bars are there? _____

5.  Sometimes the bars go across.
    How many bars are there? _____

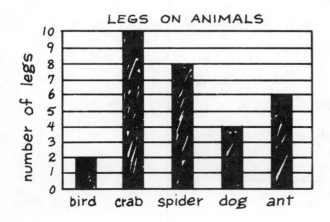

LEGS ON ANIMALS

number of legs
10 9 8 7 6 5 4 3 2 1 0

bird  crab  spider  dog  ant

POPULAR STATE BIRDS

bluebird / cardinal / robin / mockingbird

0  1  2  3  4  5  6  7  8
number of states

© 1981 by Dale Seymour Publications

27

# ERNEST'S MEMORY TEST

Ernest gave his friends one minute to memorize ten words.
After the time was up, he asked them to say the words.
He drew a bar graph to show the results.

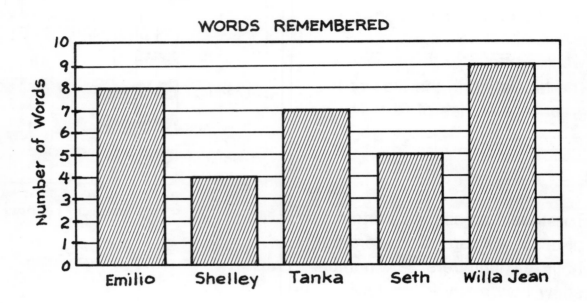

**Write *yes* or *no* in the blanks. Use Ernest's bar graph.**

1. Emilio remembered exactly 8 words. _____
2. Tanka remembered exactly 8 words. _____
3. Shelley remembered exactly 1 word. _____
4. Willa Jean remembered the most words. _____
5. Emilio remembered the most words. _____
6. Seth remembered the fewest words. _____
7. Shelley remembered the fewest words. _____
8. Willa Jean remembered more words than Seth. _____
9. Tanka remembered more words than Seth. _____
10. Shelley remembered more words than Seth. _____
11. Tanka remembered fewer words than Shelley. _____
12. Tanka remembered fewer words than Emilio. _____

Try your own memory test with friends. Use these words.

spider    caterpillar    ant    lion    tiger    bear    red    green    purple    black

# GRAPH-ITY

In each problem, only one graph shows *all* the facts correctly. Circle the correct graph. Draw an X on each of the other two graphs.

**1.** Pigs live about 10 years.
Cows live about 20 years.
Lions live about 30 years.

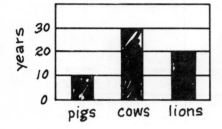
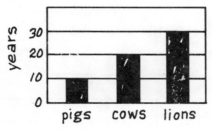
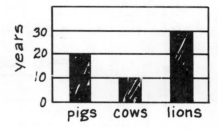

**2.** A toothbrush is about 20 cm long.  A hairbrush is about 25 cm long.
A hairpick is about 15 cm long.

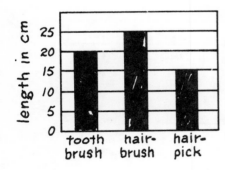
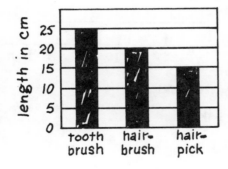
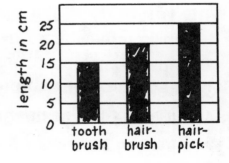

**3.** Kim is 12 years old.  Simon is 10 years old.  Sucha is 8 years old.
Lita is 10 years old.

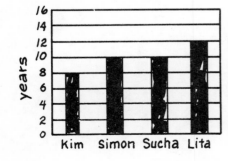
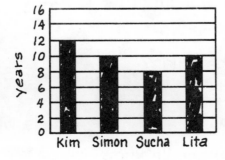
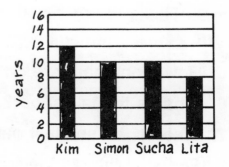

Name _____

# WHO DO YOU KNOW?

Homer asked his friends about jobs.
He kept tallies of the answers.

| Do you know someone who is | yes | no |
|---|---|---|
| a rock star ? | I | ⳾⳾⳾⳾⳾ I |
| a farmer ? | II | ⳾⳾⳾⳾⳾ |
| a typist ? | IIII | III |
| a chef ? | II | ⳾⳾⳾⳾⳾ |
| a janitor ? | ⳾⳾⳾⳾⳾ | II |
| a Martian ? | | ⳾⳾⳾⳾⳾ II |
| a carpenter ? | III | IIII |

**Complete the bar graph that Homer started.**

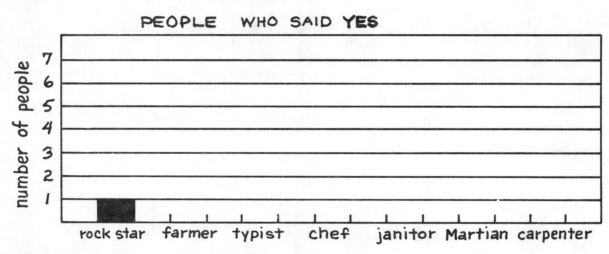

PEOPLE WHO SAID **YES**

**Solve each problem about the graph.**

1. What is the title of the graph? _____

2. Write a different title for the graph. _____

3. What do the numbers at the side of the graph tell?

   _____

4. What do the words at the bottom of the graph tell?

   _____

5. How many bars are on the graph? _____

6. What is the tallest bar? _____

7. What is the shortest bar? _____

8. For what job did the most people know someone? _____

Name _____

# SNAKES ALIVE!

Mary read that some snakes are very long.

She saw this chart.

| Snake | Length |
|---|---|
| African viper | 2 meters |
| Anaconda | 8 meters |
| Boa constrictor | 3 meters |
| Cobra | 2 meters |
| Python | 9 meters |

**Complete the bar graph to show the information about snakes.**

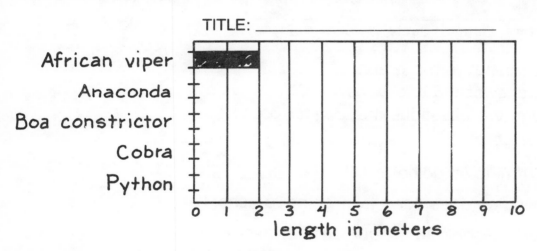

TITLE: _____

African viper
Anaconda
Boa constrictor
Cobra
Python

0   1   2   3   4   5   6   7   8   9   10
length in meters

**Answer the questions about your graph.**

1. What is the title of your graph? _____

2. How many bars are on the graph? _____

3. How long is the anaconda? _____

4. How long is the cobra? _____

5. Which bar is longest? _____

6. Which bar is shortest? _____

7. Which bar shows 3 meters long? _____

8. Which bar shows 10 meters long? _____

9. Which bar shows 5 meters long? _____

31

# TAKING TIME

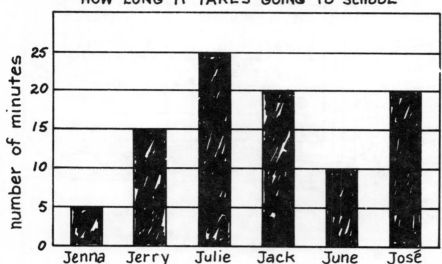

HOW LONG IT TAKES GOING TO SCHOOL

number of minutes

Jenna    Jerry    Julie    Jack    June    José

You can't tell everything from a graph.
Use the graph to answer the questions.
Draw a line through the questions that *cannot* be
answered from the graph.

1. Who takes 20 minutes going to school? _____

2. How many minutes does June take going to school? _____

3. Who takes longer going to school, Julie or Jack? _____

4. Who has the farthest to go? _____

5. How many minutes does Jim take going to school? _____

6. Is Julie late to school if she starts at 8:00 AM? _____

7. Does José get to school before Jerry? _____

8. When does Jenna get to school if she starts at 8:25? _____

9. Does June walk faster than Jenna? _____

10. How many more minutes does it take Jack than Jerry? _____

11. Who takes the longest going to school? _____

12. If Jenna, Jerry, Julie, Jack, June, and José start at the same time,
    who gets to school first? _____

Name _____

# I ♡ MY PET

Becky likes her pet rabbit very much. She asked her friends what pets they like. Each friend had one vote.

She made a tally to show their answers.

| Pet | Number of friends |
|---|---|
| rabbit | /// |
| dog | ＨＨ /// |
| cat | ＨＨ //// |
| fish | ＨＨ |
| mouse | |
| horse | // |
| other | ＨＨ /// |

Next, Becky started a graph.

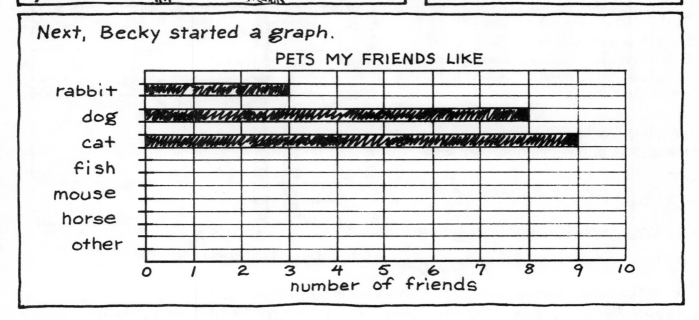

PETS MY FRIENDS LIKE

number of friends

**Complete the graph for Becky.**
**Then answer the questions about Becky's bar graph.**
**When there are not enough facts, write** *not enough facts.*

1. Which pet was the favorite? _____

2. Which pet was named least? _____

3. How many friends said *cats* or *dogs*? _____

4. How many pets were named by more than 5 friends? _____

5. How many fewer friends said *horses* than *fish*? _____

6. How many friends did *not* say *mouse*? _____

7. Did any of Becky's friends say *hamster*? _____

# COUNTING LETTERS

Count the number of letters in each word of this story. Make a tally. Then put your facts on the graph. Don't forget to count the title.

### RUNNING A RACE

One day Zippo Rabbit said to Pokey Turtle, "I'll race you to the river." "OK," said Pokey. "One, two, three, GO!"

Zippo darted off. Pokey was left far behind. But he was thinking, "I know this story! The rabbit falls asleep because she is so sure of winning. Then I win the race!"

So Pokey kept crawling along. When he got to the river, Zippo was already there. She hadn't fallen asleep. She had won.

"But that's not what happened in the story!" said Pokey.

MORAL: Don't believe everything you hear.

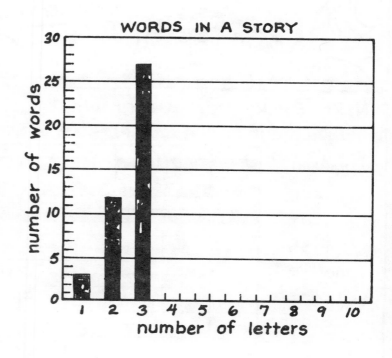

Answer each question. When there is no answer, write *none*.

1. How many words have 4 letters? _____

2. How many words have 6 letters? _____

3. How many words have 12 letters? _____

4. How many words have 6 or 7 letters? _____

5. How many words have more than 8 letters? _____

6. How many words have less than 4 letters? _____

7. How many words do *not* have 5 letters? _____

8. How many words are in the story? _____

9. How many letters are in the story? _____

Name _____

# TV TIME

Dino kept a record of the amount of time he watched TV last week.
His record looked like this.

| Sun. | Mon. | Tues. | Wed. | Thurs. | Fri. | Sat. |
|------|------|-------|------|--------|------|------|
| 3 hr | 2 hr | 1 hr | 1 hr | 2 hr | 2 hr | $2\frac{1}{2}$ hr |
| $\frac{1}{2}$ hr | | | $1\frac{1}{2}$ hr | | 1 hr | 1 hr |
| $1\frac{1}{2}$ hr | | | $\frac{1}{2}$ hr | | | 2 hr |
| | | | | | | $\frac{1}{2}$ hr |

**Complete the following about Dino's TV time.**

1. For each day find the total number of hours Dino spent watching TV.

   Sun.    Mon.    Tues.    Wed.    Thurs.    Fri.    Sat.

   _5_    ____    ____    ____    ____    ____    ____

2. Make a bar graph below to show Dino's TV watching.
   Give a title to your graph.

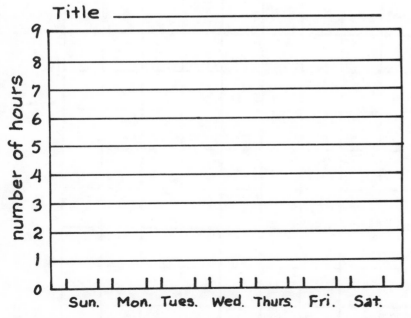

Title _____

(bar graph: y-axis "number of hours" labeled 0–9; x-axis labeled Sun. Mon. Tues. Wed. Thurs. Fri. Sat.)

Keep a record of how much time you spend watching
TV for one week. Make a graph to show your TV time.

35

# A HEAVY PROBLEM

Tony and Nugyen plan to make a bar graph. They will use the information at the right.

| baseball | 150 grams |
|---|---|
| basketball | 600 grams |
| football | 400 grams |
| soccer ball | 500 grams |
| tennis ball | 60 grams |

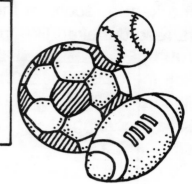

**Answer these questions to help plan the graph.**

1. What title should they give the graph? _____

2. How many bars will they make? _____

3. What is the heaviest ball? _____

4. They want the gram measurements by 50s across the bottom of their graph. How many lines do they need? _____

On the graph paper below, make a bar graph for Tony and Nugyen. Don't forget to make a title and label the parts.

Use dice.

Name _____

# WHAT'S THE MAGIC NUMBER?

This is an experiment.
You need a pair of dice.
Roll the dice 20 times.
Record the sums by shading squares.
Start recording at the bottom of the chart.

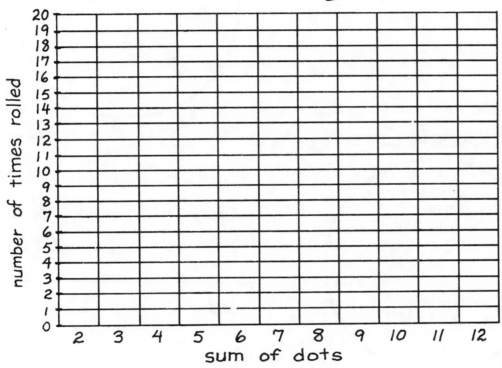

**Solve each problem.**

1. Which sum did you see most often? _____

2. Which sum did you see least often? _____

3. Suppose you roll the dice 40 times.

   About how many times would you see 4 dots? _____

4. Suppose you roll the dice 40 times.

   About how many times would you see 7 dots? _____

5. Suppose you roll the dice one more time.

   How many dots are you most likely to see? _____

# CIRCLE GRAPHS

**Circle graphs** compare parts of a whole.

Circle graphs have **titles.**

1. The title of this circle graph is

_____.

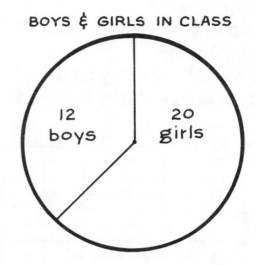

BOYS & GIRLS IN CLASS

12 boys    20 girls

GLASSES IN CLASS

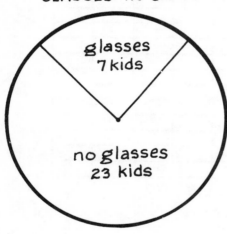

glasses
7 kids

no glasses
23 kids

Each part of a circle graph has a **label.**

2. There are 2 parts to this circle graph.
The labels for each part are

_____,

and _____.

A circle graph is separated into different parts like the slices of a pie. The biggest slice means the greatest part. The smallest slice means the least part.

3. This graph has _____ parts.

_____ hair is the biggest part.

_____ hair is the smallest part.

HAIR COLOR IN CLASS

blonde hair
8 kids

black hair
5 kids

red hair
2 kids

brown hair
20 kids

Name _____

# CHOOSE THE GRAPH

In each problem, only one graph shows *all* the
facts correctly. Circle the correct graph. Draw
an X on each of the other two graphs.

1. There are 10 cousins in all. Dino has 5 cousins.
   Ellie has 2 cousins. Josie has 3 cousins.

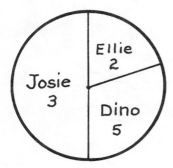
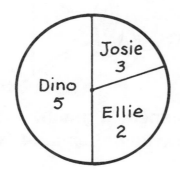
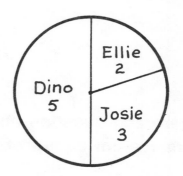

2. A pizza was cut into 6 equal pieces. Jonah ate 2 pieces.
   Peter ate 3 pieces. Mary Jane ate 1 piece.

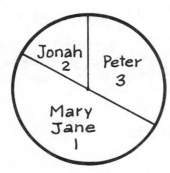
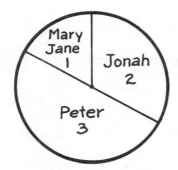
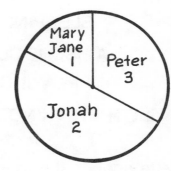

3. Yetta had 25 cents. Ric had 5 cents. Jolene had 10 cents.
   They had 40 cents altogether.

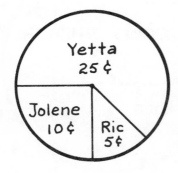
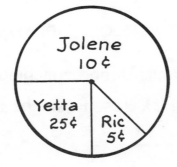
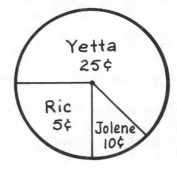

39

Name _____

# BUSY DAY

Janetta made a circle graph to show how she spends her time in one day.

HOW I SPEND MY DAY
by Janetta Johnson

eat 2 hr

other things 5 hr

play 3 hr

school 6 hr

sleep 8 hr

**Answer each question about Janetta's graph.**
**If there are not enough facts, write *not enough facts*.**

1. How many hours does Janetta spend on each thing?
   Write your answers on the blanks.

   EATING _____

   SLEEPING _____

   AT SCHOOL _____

   PLAYING _____

   DOING OTHER THINGS _____

There are 24 hours in a day. Subtract to find this answer.

2. Janetta spends the most time on which thing? Circle one.

   EATING      SLEEPING      AT SCHOOL      PLAYING      DOING OTHER THINGS

3. Janetta spends the least time on which thing? Circle one.

   EATING      SLEEPING      AT SCHOOL      PLAYING      DOING OTHER THINGS

4. How many hours does Janetta spend watching TV? _____

5. How many hours do you think Janetta sleeps altogether

   in seven days? _____

40

# ABOUT LINE GRAPHS

**Line graphs** help show changes.

Line graphs have **titles**.

1. The title of this line graph is

_____.

2. The labels along the side of this

graph are numbers for _____.

The numbers start at _____ and end

at _____.

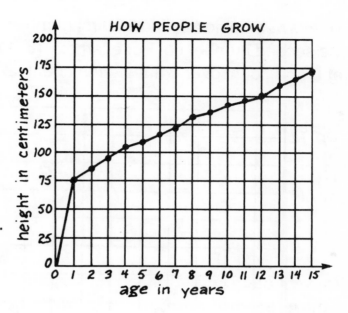

3. The labels across the bottom of

this graph are numbers for _____.

The numbers start at _____ and end

at _____.

Line graphs use a **line** to show how things change.
The line goes up, goes down, or stays the same.

4. The line on this graph goes _____

from 10 A.M. to 1 P.M. It _____

from 1 P.M. to 4 P.M. It _____

from 4 P.M. to 6 P.M.

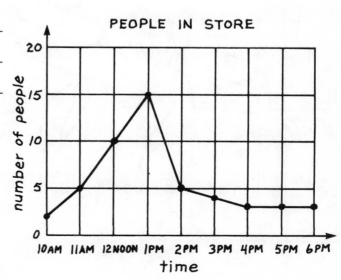

41

# GET THE POINT?

To answer the riddle, find the first number over and
the second number up. For example, (8, 11) means start
at 0 and count *over* 8 lines, then count *up* 11 lines.

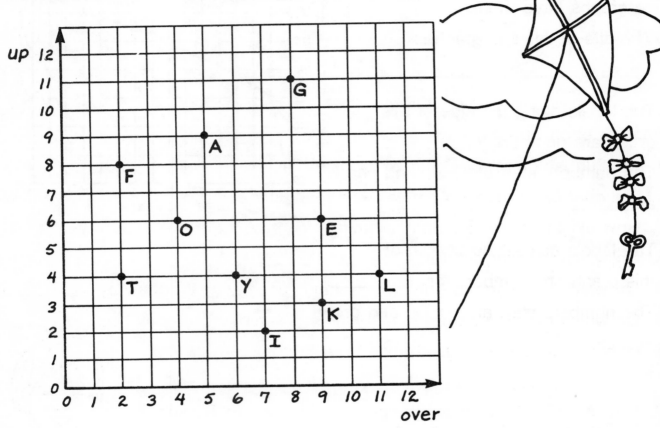

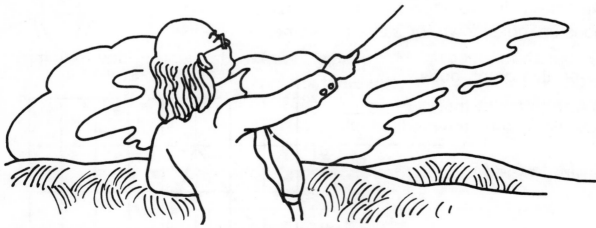

Ben Franklin's friend said this about his ideas.

<u> G </u>  __  __  __  __  __  __  __  __  __!
(8,11) (4,6)  (2,8) (11,4) (6,4)  (5,9)  (9,3) (7,2) (2,4) (9,6)

42

Name _____

# UP OR DOWN?

Erica drew a graph to show how her feelings changed last week.
Her graph looked like this.

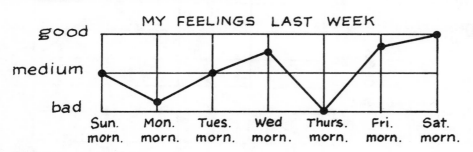

**Answer each question about Erica's graph.**

1. On Sunday morning, Erica felt medium.

   On what other morning did she feel medium? _____

2. How did Erica feel Saturday morning? _____

3. On which morning did Erica feel bad? _____ __

4. On which morning did Erica feel better, Tuesday or Wednesday? _____

5. On which morning did Erica feel worse, Thursday or Friday? _____

6. How did Erica's feelings change from Sunday to Monday?

   _____

7. How did Erica's feelings change from Monday to Tuesday?

   _____

Last week, Jonah felt medium on Tuesday, Wednesday, and Friday morning.
He felt good on Thursday and Saturday mornings.
He felt bad on Monday morning.
On Sunday morning he felt medium bad.

**Draw a graph to show how Jonah's feelings changed last week.**

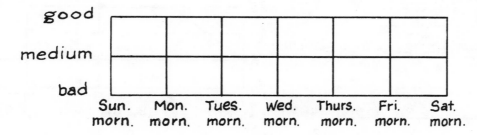

# WHAT'S MY LINE?

Rita is trying to get better at bowling. She is keeping a line graph to show her averages.

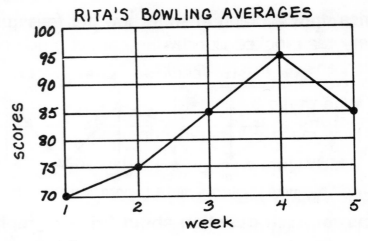

RITA'S BOWLING AVERAGES

**Complete each of the following.**

1. What was Rita's average the first week? _____

2. What was Rita's average the second week? _____

3. What was Rita's average the third week? _____

4. What was Rita's average the fourth week? _____

5. What was Rita's average the fifth week? _____

6. How did Rita's average change? Write *better* or *worse* on the blanks.

   from week 1 to week 2 _____

   from week 2 to week 3 _____

   from week 3 to week 4 _____

   from week 4 to week 5 _____

7. The line graph at the right shows Adam's averages. Tell how his averages changed from week to week.

   _____

   _____

   _____

   _____

   _____

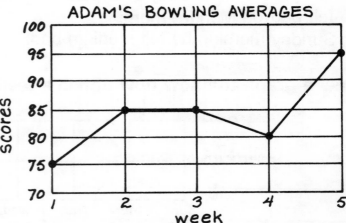

ADAM'S BOWLING AVERAGES

Name _____

# STRIKE A MATCH

Each graph or table answers two of these questions.
Write the questions under the correct graph or table.
Then answer the questions.

1. Who made 3 shots?

2. How many tickets did Anna sell?

3. Which team won the most games?

4. How many people played?

5. How much did Jake spend?

6. What is the key?

7. Which flower is least popular?

8. Who had more wins, Marks or Mains?

9. Which ones got more than five votes?

### FREE THROW CONTEST

| name | shots made |
|------|-----------|
| Inez | 8 |
| Hori | 3 |
| Albert | 5 |
| Elaine | 7 |

_____

_____

### TICKETS SOLD

Dale ▭
J.B. ▭ ▭ ▭ ▭
Anna ▭
Gail ▭ ▭ ▭

▭ means 2 tickets

_____

_____

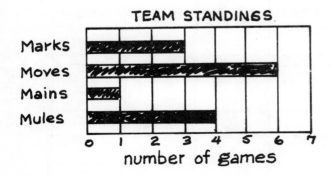

TEAM STANDINGS

Marks, Moves, Mains, Mules — number of games

_____

_____

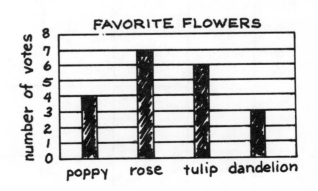

FAVORITE FLOWERS

number of votes — poppy, rose, tulip, dandelion

_____

_____

© 1981 by Dale Seymour Publications

45

Title: _____

Name _____

Title: _____

Name _____

# ANSWERS

## page 1

1. first column: 5, 10, 15, 20;
   second column: 8, 13, 18, 23
2. first column: ///, ////, //////// //, //// //;
   second column: //// //// //// ///, //// //// //// //// ////,
   //// //// //// //// //// /, //// //// //// //// /

## page 2

1. blue, brown, green
2. //// // , 7
3. 7
4. 19

5. 19
6. 3
7. 3
8. 29

## page 3

1.

| digit | number of times |
|-------|-----------------|
| 1 | // |
| 3 | / |
| 5 | /// |
| 6 | / |

2.

| digit | number of times |
|-------|-----------------|
| 2 | / |
| 5 | // |
| 6 | / |
| 7 | // |
| 9 | / |

3. Answers will vary.
   Be sure the tallies accurately
   represent the number of digits
   in the phone number given.

## page 4

1.

| digit | number of times |
|-------|-----------------|
| 0 | /// |
| 1 | //// //// //// |
| 2 | //// //// /// |
| 3 | //// |
| 4 | /// |
| 5 | /// |
| 6 | /// |
| 7 | /// |
| 8 | /// |
| 9 | /// |

2. 1

3.

| digit | number of times |
|-------|-----------------|
| 0 | /// |
| 1 | //// //// //// |
| 2 | //// //// /// |
| 3 | //// |
| 4 | /// |
| 5 | /// |
| 6 | /// |
| 7 | /// |
| 8 | /// |
| 9 | /// |

4. 1
5. yes

## page 5

1. 12
2. 2
3. 3
4. brown
5. blue

6. white
7. 33
8. 21
9. 26

## page 6

Answers will vary. Be sure the answers
to problems **1-5** accurately represent
the data given in the tally chart.

## page 7

1 Number of Players on Team
2. name, shots made
3. Lions, Eagles, Bears, Roosters
4. 7, 8, 10, 8

## page 8

1. Monday
2. Friday
3. Mom's birthday
4. Trip to museum

5. February 16
6. February 19
7. Thursday
8. 5

## page 9

1. 25
2. 40
3. 20
4. 20
Anna (by 25 sit-ups)

5. Lee
6. Anna
7. Ray
8. Ray and Anna

**page 10**

Title: Exercises Kids Did (answers will vary)

| Kid's Names | jumping jacks | sit-ups | touch-toes |
|---|---|---|---|
| Jackie | 10 | 15 | 12 |
| Otis | 20 | 20 | 20 |
| Gretchen | 25 | 10 | 15 |
| Ben | 15 | 25 | 10 |

**1.** 70    **6.** 35
**2.** 70    **7.** Gretchen
**3.** 57    **8.** Ben
**4.** 35    **9.** Otis
**5.** 25

**page 11**

Starting with Anchorage, the temperature changes are 24°, 22°, 30°, 37°, 15°, 30°, 28°, 21°, 16°.

**1.** San Juan    **4.** Houston
**2.** Anchorage    **5.** San Juan
**3.** Detroit    **6.** Anchorage

**page 12**

**1.** 6
**2.** The Tick-Tock Man
**3.** The Monster with Three Eyes
**4.** page 9
**5.** page 15
**6.** page 17
**7.** page 8
**8.** 4 pages
**9.** page 16
**10.** 2 pages

**page 13**

**1.** 7:59 AM    **4.** no, no, yes
**2.** 8:32 AM    **5.** 23 minutes
**3.** 9:01 AM

**page 14**

**1.** yes    **6.** yes
**2.** no    **7.** yes
**3.** no    **8.** yes
**4.** yes    **9.** no
**5.** yes    **10.** yes

**page 15**

Answers will vary.

**page 16**

Answers will vary.

**page 17**

**1.** Milk Sales
**2.** Dale, J.B., Anna, Gail
**3.** trees
**4.** 2 mittens

**page 18**

**1.** Thursday
**2.** Wednesday
**3.**

NUMBER OF BIRTHDAYS

| | | | | | | | | | |
|---|---|---|---|---|---|---|---|---|---|
| Sunday | ■ | ■ | ■ | ■ | | | | | |
| Monday | ■ | ■ | ■ | ■ | ■ | | | | |
| Tuesday | ■ | ■ | ■ | | | | | | |
| Wednesday | ■ | ■ | | | | | | | |
| Thursday | ■ | ■ | ■ | ■ | ■ | ■ | ■ | ■ | |
| Friday | ■ | ■ | ■ | ■ | | | | | |
| Saturday | ■ | ■ | | | | | | | |

**page 19**

**1.** 5    **7.** 0
**2.** 1    **8.** 2
**3.** 3    **9.** 3
**4.** 8    **10.** 4
**5.** 2    **11.** 5
**6.** 4    **12.** 23

Questions and answers will vary.

**page 20**

**1.** Favorite Colors    **7.** red
**2.** 9    **8.** orange and purple
**3.** 1 person    **9.** green and brown
**4.** blue    **10.** yellow and black
**5.** white    **11.** 30
**6.** yellow    **12.** 24

**page 21**

**1.** desk, 3 feet;
chair, 2 feet;
chalkboard, 7 feet;
door, 3 feet
**2.** 3 feet, desk and door;
2 feet, chair;
7 feet, chalkboard
**3.** desk, chalkboard, chalkboard, chalkboard
**4.** chair, chair, door, chair
**5.** desk and door
**6.** chalkboard, door or desk, chair
**7.** the second picture is correct

**page 22**

1. the third picture is correct
2. the third picture is correct
3. 80¢
4. $1.80
5. 10¢ (the food cost $1.90)

**page 23**   TITLE: _Cost of Toys at the Store_

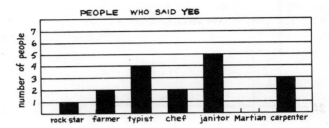

(10¢) means 10 cents

1. Cost of Toys at the Store
   (compare this answer to the title)
2. 10 cents
3. 8
4. show half of a coin
5. show $3\frac{1}{2}$ coins

**page 24**

1. b        4. c
2. d        5. a
3. c        6. d

**page 25**

1. Hitown              4. Lowtown
2. 3,000               5. 3,000
3. Intown and Hitown   6. 2,000

**page 26**

1. picnicing, driving, swimming,
   playing games, and boating
2. starting with the first picture:
   20,000,000;
   40,000,000;
   60,000,000;
   120,000,000;
   10,000,000;
   50,000,000
3. 160,000,000
4. 100,000,000
5. 50,000,000

**page 27**

1. Speeds of Swimmers
2. trout, barracuda, whale, dolphin
3. 0, 50
4. 5
5. 4

**page 28**

1. yes        7. yes
2. no         8. yes
3. no         9. yes
4. yes        10. no
5. no         11. no
6. no         12. yes

**page 29**

1. There should be X's on the first and
   last graphs, and a circle around the
   middle graph.
2. There should be X's on the second and
   third graphs, and a circle around the
   first graph.
3. There should be X's on the first and
   last graphs, and a circle around the
   middle graph.

**page 30**

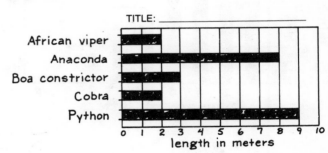

1. People Who Said Yes        5. 7
2. Answers will vary.         6. janitor
3. how many people           7. Martian
4. different jobs            8. janitor

**page 31**

TITLE: _____

1. Answers will vary.      6. African viper and cobra
2. 5                       7. boa constrictor
3. 8 meters                8. none
4. 2 meters                9. none
5. python

**page 32**

1. Jack and José
2. 10 minutes
3. Julie
4. question cannot be answered
5. question cannot be answered
6. question cannot be answered
7. question cannot be answered
8. 8:30
9. question cannot be answered
10. 5
11. Julie
12. Jenna

**page 33**

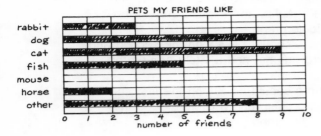

PETS MY FRIENDS LIKE

1. cat
2. mouse
3. 17
4. 3
5. 3
6. 35
7. no enough facts

**page 34**

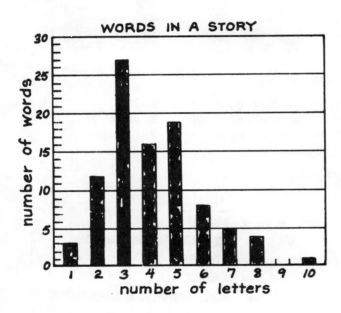

WORDS IN A STORY

1. 16
2. 8
3. 0
4. 13
5. 1
6. 42
7. 75
8. 94
9. 384

**page 35**

1. Sun. Mon. Tues. Wed. Thurs. Fri. Sat.
   5    2    1    3    2    3    6

2.
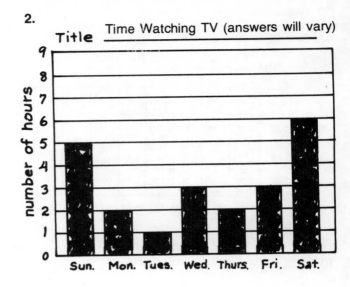

Title _Time Watching TV (answers will vary)_

**page 36**

1. How Many Grams in Game Balls
   (Answers will vary.)
2. 5
3. basketball
4. 13 if they start at 0
   Graphs will vary.

**page 37**

Answers will vary. Be sure the answers accurately represent the
data recorded in the graph.
The *most likely* answers are:
1. 7
2. 2 or 12
3. 4 or 5
4. 6 or 7
5. 7

**page 38**

1. Boys and Girls in Class
2. glasses, 7 kids;
   no glasses, 23 kids
3. 4; brown, red

**page 39**

1. The first two graphs should have X's
   on them. The last graph should be circled.
2. The first and last graphs should
   have X's on them. The middle graph
   should be circled.
3. The first graph should be circled.
   The other two should have X's on them.

**page 40**

1. eating, 2 hr;
   sleeping, 8 hr;
   at school, 6 hr;
   playing, 3 hr;
   doing other things, 5 hr
2. sleeping
3. eating
4. don't know
5. 56 hr

**page 41**

1. How People Grow
2. height in centimeters, 0, 200
3. age in years, 0, 15
4. up, goes down, stays the same

**page 42**

GO FLY A KITE!

**page 43**

1. Tuesday
2. good
3. Thursday
4. Wednesday
5. Thursday
6. They got worse.
7. They got better.

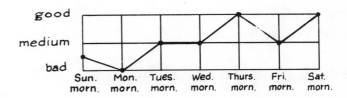

**page 44**

1. 70
2. 75
3. 85
4. 95
5. 85
6. better, better, better, worse
7. Answers will vary.

**page 45**

Free Throw Contest, 1 and 4
Tickets Sold, 2 and 6
Team Standings, 3 and 8
Favorite Flowers, 7 and 9